Mendenhall Glacier
· FLOWING THROUGH TIME ·

By Katherine Hocker

Photography by David Job

Introduction

Over thousands of years, the ice-river that we call the Mendenhall Glacier has surged and receded: sculpting and building, destroying, refining. Every cliff, every speckled boulder, every smoothed outcrop within its valley recalls the slow power of the ice. Every stone tells the Mendenhall's story.

The new life that emerges at the face of the receding glacier knows that story, and responds to it. Chalky gray lichens creep their patient way across the scoured rocks, while nearby, lupines bloom from more generous glacial soil. Below, blood-red sockeye salmon find shelter in ponds where icebergs once lay.

We respond to the glacier, too. Hundreds of thousands of us—residents and tourists alike—visit this glacier each year. We take pictures of it, explore its crevasses, touch its icebergs, study its movements, wade in the frigid lake waters.

Why do we come? Maybe it's to admire the sterile beauty of the ice-river, juxtaposed against the abundance of the temperate rainforest. Maybe it's to get a glimpse into an arctic world that few of us will ever enter. Or maybe it's to encounter, explore, and appreciate—just once—a force that is truly bigger than us all.

1

A Land of Glaciers

Glaciers are our past, present, and future; they shape our lives in more ways than we immediately recognize. Glaciers cover 10 percent of the Earth's surface, and store over 75 percent of its fresh water. By exposing the Bering Land Bridge, the great Pleistocene glaciers made possible the migration of humans from Asia to North America tens of thousands of years ago. Some of our richest farmlands began as deep drifts of glacial loess, or dust. Glaciers have shaped some of our most beloved natural wonders, including Yosemite Valley and the Matterhorn.

Alaska is home to the largest glaciers in the world (outside of the Antarctic and Greenland ice caps), and the southern coast of Alaska is particularly rich in glacier ice. Among the rugged mountains of the Coast Range, the Fairweather Range, and the Wrangell-St. Elias Mountains, glaciers are born by the hundreds. The Mendenhall Glacier is one of the most well known of these coastal glaciers; it flows from the 1,500-square-mile Juneau Icefield.

The Juneau Icefield is one of the largest icefields in North America. Meltwater from its glaciers feeds both coastal and interior rivers—including the great Yukon River.

Glacial Nurseries
Glaciers can form anywhere that annual snowfall consistently exceeds annual snowmelt. They thrive in temperate coastal climates where winter snowfalls are staggering and summer temperatures are cool. Southeast Alaska, Chile, Norway, and New Zealand are all rich in icefields and glaciers.

Water—whether liquid or ice—absorbs long wavelengths of light (reds and yellows), but allows short-wave light (blue) to pass through. Thus it acts as a filter, removing red and yellow light. That's why underwater photos appear blue. Since glacier ice is almost pure water, it has an intense sapphire color. Where there are air bubbles trapped in the ice, or where it has begun to break apart from melting, it appears white. This is why freshly calved surfaces of the glacier appear dark blue at first, then fade to white.

Water Transformed

The Mendenhall Glacier is a noisy being. Spend time in its presence, and you'll hear its many voices—from shattering crashes to bell-like chimes. The ice sizzles, groans, roars, crackles. It's rarely quiet, and never silent.

Why? Because it's a glacier—and, put most simply, a glacier is ice in motion. At a quick glance, the Mendenhall may seem just a pile of ice, enormous, but static. It's not. It's an incredibly dynamic force: noisy, changeable…and powerful enough to carve mountains.

Formation

The Mendenhall is the inevitable offspring of weather, water, and gravity, combined perfectly in the Coast Range of Southeast Alaska. Here, wet Pacific air meets and rises against the wall of 5,000 to 8,000-foot peaks. As it rises, the air cools, and as it cools, the water condenses. At sea level, this usually means rain, which nourishes the thick mossy forests that cloak the mountainsides. But at the crest of the Coast Range, it means snow, and plenty of it. In an average winter, the equivalent of over 100 feet of fresh snow may fall at high elevations.

In an inland climate, much of this snow would melt in the heat of summer. But the moist breath of the Pacific

Glacial ice is unlike any other type of ice in nature. Because it is formed from already frozen material, under huge pressure, it's extremely dense (denser than ice formed by freezing alone, and nearly as dense as liquid water). It's almost perfectly transparent. And its free-form individual ice crystals can grow to the size of soccer balls.

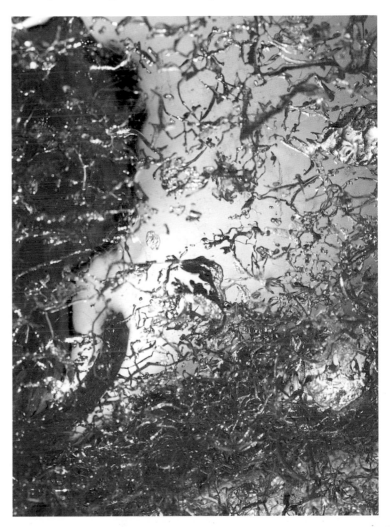

keeps Southeast Alaskan summers cool. So the snow stays… and stays… and stays, with each year's snow accumulation covering the last.

As the old snow settles downward—pulled by gravity and compressed by the weight of snow above—it begins to change. Delicate snow crystals are smoothed and squeezed together. Water melts out, trickles inward, and re-freezes, forming granular ice crystals called firn. As the pressure increases, the firn crystals merge, grow, and re-align. Air is gradually forced out from between them, leaving behind a substance that is very nearly pure ice: glacial ice.

Here in Southeast Alaska, the time it takes for a snowflake to be prodded and squeezed into glacier ice is relatively short. Much of the snow that falls on the Juneau Icefield is already wet and dense, and snow accumulates quickly. This means it takes only a few years to turn fresh snow into glacier ice. In colder, drier climates, it may take thousands of years for the same process to occur.

The Juneau Icefield has been around for thousands of years, but its glaciers flow quickly, meaning that most of their ice is fairly young. Ice at the face of the Mendenhall Glacier is only about 250 years old.

Flow

As glacier ice builds in the icefield, it begins to move downward and outward, pulled by gravity, deforming to fit the contours of the land around. The tendrils of ice that result—reaching down valleys into Alaska, and Canada's British Columbia and Yukon Territories—are the glaciers of the Juneau Icefield.

Two things cause the Mendenhall and its sister glaciers to flow. One is the natural deformation of the ice, as crystals—impelled by the weight of ice above—slip across each other and downhill. The other is water: pressure, internal friction, and heat from the ground below cause some of the ice to melt and trickle down to the glacier's base, where it forms a lubricating layer that helps the mass of ice slip along its path. Meltwater percolates through the glacier as well, adding to the slippery river below.

Recession

Despite its constant downhill flow, the Mendenhall glacier is steadily getting smaller. Tourists of the 1940s saw an ice face only a few hundred feet from the site of the present-day visitor center. By the turn of the 21st century, the glacier had receded nearly a mile.

Glaciers advance and recede according to cycles of climactic warming and cooling. The current global warming trend has meant that glacial ice in the Juneau Icefield is not being made fast enough to replace the ice that is melting away. The Mendenhall shrinks in mass nearly every year. Between 1948 and 2000, the Mendenhall lost the equivalent of a cubic mile of ice.

The Mendenhall's retreat is made speedier by ice loss at its terminus (face). The Mendenhall's broad terminus lies in a lake of its own making. At deeper spots along the terminus, the water partially lifts the glacier face, jostling it and causing it to shatter into large chunks, or icebergs. This dramatic process is called calving, and when it is happening, the glacier face recedes much more quickly.

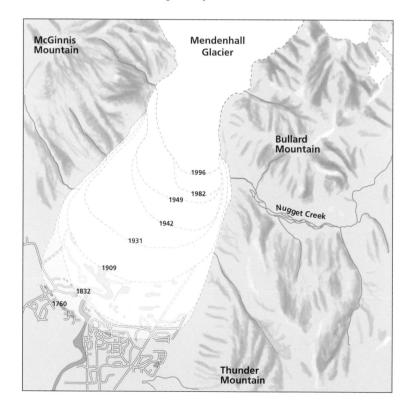

The ice at the heart of the glacier is pliable, but surface ice is brittle. Where the ice is under stress (as when the glacier flows over a hill), the base layer bends but the surface layer shatters, forming crevasses that may be over 100 feet deep.

The glacier moves fastest at its center, and slowest where its sides drag along the valley walls. The difference in speeds between the center and the sides causes stress to the ice, cracking the surface.

9

Breaker, Maker

Children love the steep-sided sandy hills near the Mendenhall Glacier Visitor Center. In winter, these are sledding hotspots, with youngsters scrambling again and again up the sides to launch for brief, but speedy, thrill rides. In summer, they're manageable mountains commanding a grand view of lake, glacier and tree-tops—just waiting to be conquered by those on small legs. The little hills are gifts of the Mendenhall Glacier, sculpted and presented in the early 1900s.

The Mendenhall, like all glaciers, is a rockbreaker: a quarrier, a polisher, an eater of bedrock. But it is also a creator: a maker of hills; a con-veyor of stones, sand, and silt; a builder of rivers and lakes. Its actions, and the movements of the water that melts from it, have carved and built the Mendenhall Valley into what it is today.

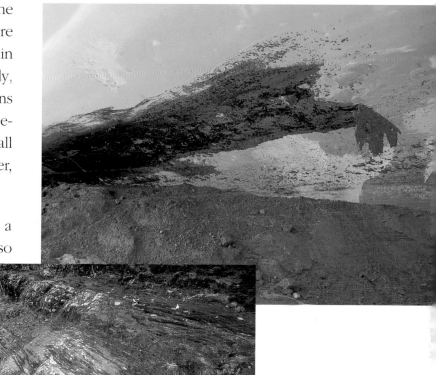

Rockbreaker

As the Mendenhall moves down-valley, meltwater at the interface between ice and rock flows into the cracks in the rock, then re-freezes, expanding and popping pieces loose. The bedrock chunks, caught up by the ice, form a layer of grit at the base of the glacier. The motion of the flowing glacier then drags

Polished bedrock surrounds the Mendenhall Glacier Visitor Center. Some rocks have deep gouges, or striations, where large stones, like those in the ice in the photo above, were dragged across them.

this "sandpaper" layer over the bedrock below, gouging, carving, polishing. It's a relentless process, and it carves spectacular large-scale landforms.

Landscaper

Balancing its talent for breaking, the glacier builds as well, using as raw material the debris that it has quarried and gathered during its journey. Boulders, cobbles, gravel, sand, and silt (together called glacial till) travel within the glacier and eventually emerge—to be shaped by ice or meltwater into unique glacial landforms.

Where the glacier meets a mountain peak, it divides to flow around; creating a mountain island called a nunatak.

Glacier-carved valleys are "U"-shaped: broad and round-bottomed, with steep sides—as opposed to the "V"-shapes of river-carved valleys.

Two glaciers flowing together create a medial moraine from their joined lateral moraines.

Lateral moraines form along the glacier's margins, and are made of material scraped and gathered from the valley walls.

From its source to its face, the Mendenhall stretches some 13 miles. At its highest elevation, its surface is almost 5,000 feet above sea level, while its face rests only about 100 feet above the sea. In some places, the ice measures well over 1,000 feet thick.

The glacier's rock-grinding produces a fine rock powder known as glacial flour, which gives the lake and river their milky appearance.

Kettle ponds began as large chunks of ice, abandoned by the receding glacier. As they melted, they left behind water-filled depressions in the outwash plain.

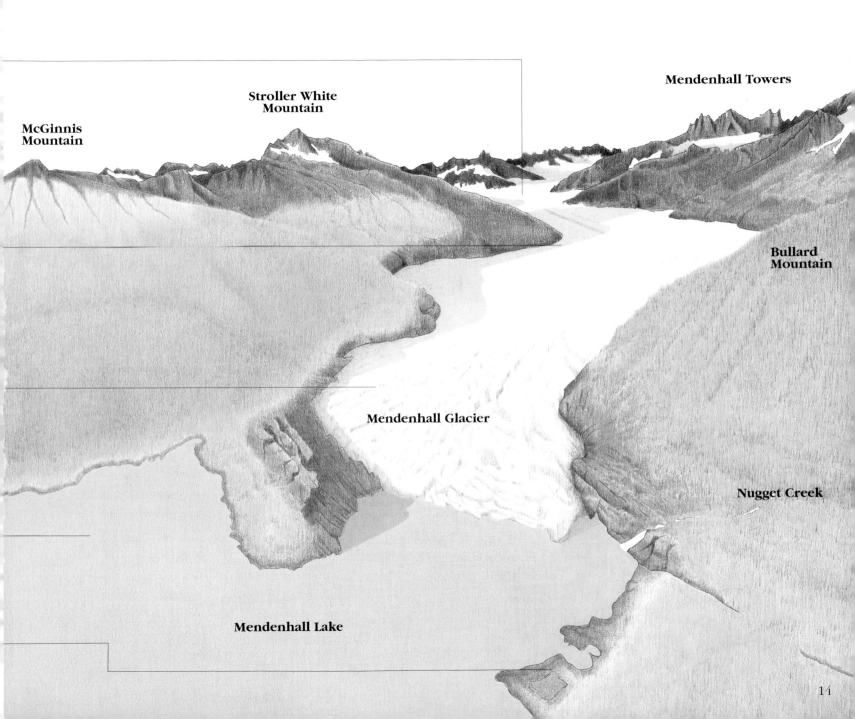

McGinnis
Mountain

Stroller White
Mountain

Mendenhall Towers

Bullard
Mountain

Mendenhall Glacier

Nugget Creek

Mendenhall Lake

14

Ice and Time

Throughout the Mendenhall Valley you'll find them: huge boulders of speckled granitic rock, some as large as trucks, resting serenely in mossy groves of spruce or sitting obstinately in lawns, grudgingly included in the landscaping. Such rocks are not native to the Valley—in fact, the closest granitic bedrock is many miles to the north, high in the Coast Range. To find the source of these stone outlanders, look to the glacier, which brought them down from the mountains hundreds of years ago.

The Mendenhall Valley is testament to its own glacial past. Ancient beach terraces, serpentine moraines, smoothed mountaintops, buried forests…and the previously mentioned boulders (known as erratics) are among the many clues that help us to piece together the glacial history of the Valley.

Sea Level Rise
10,000 years BP

As the great Ice Age ended, the ice melted, and sea levels rose. The Mendenhall Valley became a huge bay, at least 500 feet deep.

How do we know?
Ancient beach terraces, with wave-cut faces and even fossil shells, have been found high on the Valley's walls.

15,000   2,000   5,000

Ice Age
20,000 years BP

Glacial ice covered nearly all of Southeast Alaska, as it did many other regions of North America and Europe. Here in the Mendenhall Valley, the ice was at least 3,000 feet thick. This, the most recent of a series of enormous worldwide glacial advances, is known as the Wisconsin Ice Age.

How do we know?
Mountaintops give us the clues: rounded peaks such as 2,800-foot Thunder Mountain ridge were overridden and smoothed by the ice, while jagged peaks, such as 4,200-foot Bullard Mountain, protruded from the ice sheet as nunataks.

Land Rise
10,000-3,000 years BP

With the weight of the ice gone, the land rebounded, turning the bay into a low, flat valley. The receding glacier filled the valley with till. A forest of spruce and hemlock, similar to present-day forests, grew on the valley floor.

How do we know?
Cutbanks on the Mendenhall River reveal stumps and broken trunks from this ancient forest, buried under more recent glacial till.

Ice Advance
1250 AD-circa 1770 AD

Cooling climate trends caused the Juneau Icefield—and hundreds of other glaciers around the world—to grow. The Mendenhall traveled downvalley to approximately the location of today's Taku Boulevard, bulldozing forests and burying the land under millions of tons of till.

How do we know?
The terminal moraine for this "Little Ice Age" advance is clearly visible from the air; trees on this moraine date no older than 1770, meaning that the moraine was created around then. Abandoned watercourses (once the drainage channels for the glacier's meltwater), meander across the valley floor below this terminal moraine.

Ice Retreat
Circa 1770 AD to Present

Global warming turns the course of the Little Ice Age advance. The Mendenhall, like most of the other advancing glaciers, begins to recede and continues to the present day.

How do we know?
Recessional moraines (dated by tree-ring analysis) mark brief pauses in the glacier's retreat. A clear "trimline" divides young vegetation (established since 1770) from older vegetation (established before 1770) on the sides of McGinnis and Bullard Mountains, showing the height to which the glacier scraped the mountainsides bare of trees. Scientists continue to record the steady shrinkage of the Mendenhall.

Birth of a Lake
1920s-1930s

The receding face of the Mendenhall exposes a deep basin, in which meltwater collects, forming Mendenhall Lake. The lake acts as a sediment trap, stopping the active glacial outwash that created much of the Mendenhall Valley. The only outlet for the glacier's meltwater is now the Mendenhall River, which cuts a channel deep into the ancient glacial and marine sediments.

How do we know?
Historic photographs and maps record the emergence of the lake and the growth of the Mendenhall River.

Land Rises Again
Circa 1770 AD to Present

The enormous weight of the Little Ice Age glaciers melts off of the land, allowing it to rise again. The land area of the Mendenhall Valley expands continuously as old glacial outwash deltas are lifted from the sea.

How do we know?
Scientists, using GPS equipment, are able to precisely measure the current rate of uplift—it's just over half an inch per year in the Juneau area. Tidal measurements dating back to 1911 record continuous uplift. Pre-1911 uplift, dating back to the 1770s, is shown by the age structure of coastal spruce trees: as you move upward from present-day sea level, you encounter successively older trees.

19

Ice and Life

From a distance, it's a landscape as barren-looking as the plains of the moon: a dark boulder and a scattering of granite pebbles, resting in a small bedrock depression. Close by looms the great mud-streaked bulk of the glacier's face. But a second glance at the scene reveals a rosette of lichen, a patch of moss, a bright flower… and a speckled plover chick—all in the blue shadow of the ice. Fed by frigid water and nearly sterile glacial soil, and nourished by its own determination, life is already moving in.

The fact that the glaciers continually create and expose new land to be colonized means that many habitats that would otherwise not exist here are able to persist—leading to a greater diversity of plant and animal life.

Renewers

Glaciers are the great renewers of the Southeast Alaskan ecosystem. When they advance, they destroy all life in their path—smashing forests, choking rivers and lakes with sediment, obliterating thousands of years of plant growth and habitat development. When they recede, they reveal a land scraped down to its sterile minimum, with little to offer for its previous tenants.

But Southeast Alaska's lush climate leaves few areas barren for long.

Within only a few years after the "destroying" glaciers recede, life is already building on itself. Seeds float in on the wind. Salmon push their way up silty creeks. At the Mendenhall Glacier, we can see this process clearly, as the post-glacial moonscape is transformed from barren plain to a mosaic of forest, pond, stream, marsh, and meadow in a process known as post-glacial succession.

Pioneers

The first colonists to thrive on the glacier-scoured land are lichens and mosses, whose spores are blown in or brought inadvertently by roaming animals. Within only a few years, these early settlers are joined by seedlings of spruce, alder, cottonwood and willow, as well as flowers such as dwarf fireweed, Nootka lupine, and purple mountain saxifrage.

Thicket Game

The first trees to sprout on well-drained deglaciated land are usually stunted, and their leaves or needles are often faded yellow, showing lack of sufficient nutrients. But as the plant community develops, trees begin to thrive and grow taller. The turning point is the introduction of nitrogen to the soil, made possible by symbiosis between plants, such as alder and lupine, and nitrogen-fixing bacteria that form nodules on their roots. Once the nitrogen deficiency has been overcome, alder, cottonwood, and willow shoot up to form dense thickets.

Succession
On extremely dry sites, such as deep moraines where the water table is particularly low, alder/willow thickets never develop. Instead, the mats of pioneering lichen, moss, and lupine gradually grow over with cottonwood and spruce trees.

Wet areas, such as kettle ponds and poorly drained depressions in the outwash plain, are colonized by horsetail, sedge, and mosses. In time, they will become small lakes or marshes—home to fish, toads, and aquatic birds.

Some migratory songbirds including American redstart, northern waterthrush, yellow warbler, and warbling vireo, nest and forage almost exclusively in post-glacial thickets.

The Dark Woods

As alder, willow, and cottonwood are working their way toward the sky, slower-growing spruce lags behind in the shade. But over time, spruces will eventually overtop their deciduous neighbors. The alder and willow trees, unable to survive in the spruces' deep shade, die. The forest floor grows a carpet of moss.

The Future

As the spruce forest ages, some of the overstory trees die, leaving gaps in the canopy. More light means that early blueberry, five-leaf bramble, Western hemlock and other plants can move into the understory, attracting more animals. Over the course of centuries, hemlock becomes the dominant tree species and there are trees of all ages represented in the old-growth forest.

Tall trees offer convenient resting and surveying perches for bald eagles and other birds.

Porcupines (above) and red squirrels are two of the few species of mammals to thrive in the thick spruce forest.

Mendenhall Lake appeared at the face of the receding glacier in the 1920s. As the glacier receded and the lake grew, sockeye (red) and coho (silver) salmon migrated up the Mendenhall River and into the lake. Both species of fish, whose young depend on lakes for survival, quickly established populations in the surrounding creeks, becoming popular attractions for fish-watching tourists.

Beavers arrived in the mid-20th century, and their presence increased in the 1980s and 1990s. Their ponds created excellent habitat for juvenile coho salmon, enhancing local populations.

As salmon populations grew, bears moved in. Today, both black and brown bears are commonly seen in the Mendenhall Lake area, and Mendenhall Glacier Visitor Center is gaining a reputation as a bear viewing site. Black bears crowd the banks of Steep Creek during the peak of the sockeye run in July, and stay through the coho run in September and October. Brown bears have been sighted during the coho run. Both black and brown bears den on the mountains surrounding the Mendenhall Valley.

Heart of Ice

For Juneau residents today, "the glacier" is a constant, powerful presence—a reminder of the extraordinary setting of their lives. From just about anywhere in the Mendenhall Valley, you can look up from your task or conversation and take a sharp breath, surprised anew at the size and shine of the ice-river, and the magnificence of the surrounding peaks.

People have lived for centuries in the breathtaking presence of Mendenhall Glacier and the surrounding peaks.

Tlingit

For centuries, the Tlingit people—original inhabitants of this glacier-carved land—have watched the glaciers ebb and flow. What we now call the Mendenhall Glacier was well known to the Aak'w kwaan people, Tlingit native Americans who had a winter village at nearby Auke Bay. They hunted, gathered edible and medicinal plants, and set fish traps in the glacier's valley. In summer, they moved to the mouth of Gold Creek (site of present-day downtown Juneau), where they caught, dried, and smoked the salmon that were their dietary staple.

To these people, the ice-river (known as *sit* in Tlingit) was a living being. It could be spoken to, conversed with, called to flow forward or asked to retreat, but it always had to be treated with deep respect—a tradition that still holds true today.

Early Explorations

Naturalist, explorer, and writer John Muir saw and noted the glacier in 1879, on a grand voyage of glacial discovery along the Southeast Alaskan coast. Muir named it the Auke Glacier, and called it "one of the most beautiful of all the coastal glaciers." The glacier's present-day name is in honor of Thomas C Mendenhall, superintendent of the U.S. Coast and Geodetic Survey during the surveying of the boundary between Canada and Alaska in the late 1880s and early 1890s.

Settlement

Euro-American settlement of the Juneau area began in the 1880s, at the mouth of Gold Creek, some 12 miles southeast of the Mendenhall Glacier. As the mining town grew, it expanded—eventually including the Mendenhall Valley. By the 1930s, there were a number of residences and businesses within sight of the glacier. Later, a dairy farm, fox farms, a few market gardens, fish camps, and small homesteads dotted the forested valley.

A post-war boom led to the large-scale settlement of the Mendenhall Valley beginning in the 1940s. The level, well-drained glacial soil made excellent building sites, and soon suburbs grew to replace the spruce forests on the old moraines. Today the Mendenhall Valley is home to thousands of people—most of whom are more than a little proud to have a glacier in their backyards.

Tourism/Recreation

Even as early as 1910 the Mendenhall was a tourist attraction. Visitors braved a tooth-rattling wagon trip over 10 miles of rough road to clamber on the rocks and ice. By the 1930s, the Civilian Conservation Corps had built a campground and stone shelter ("Skaters Cabin") on the west side of the lake and a stone shelter on the Trail of Time, and local beauties danced the "Iceworm Wiggle" across the moraines, to the delight of visiting conventioneers. The present-day Glacier Visitor Center (named *Sit' Ya*, or "facing the glacier" in Tlingit) was built in 1962 and extensively remodeled between 1997 and 1999.

Today, the Mendenhall Glacier Recreation Area (managed by the U.S. Forest Service) sees over 350,000 visitors each summer, 95 percent of whom are from outside Juneau. In the winter, the percentages are reversed as locals come to ski, skate, hike, and ice-climb at their "backyard" glacier.

Research/Education

Dynamic and accessible, the Mendenhall Glacier is an outstanding natural laboratory for scientific research. Many significant advances in glaciology, biology, and climatology have been made by scientists working on or around the glacier. The University of Idaho's Juneau Icefield Research Program, in place since 1948, has made the Juneau Icefield one of the most thoroughly studied in the world. Each summer, scores of students and professors move into semi-permanent field camps high on the icefield to record glacier movement, study the botany of the nunataks, and peruse the layers of ice within the crevasses.

What will the Mendenhall do in the future? If the current global warming trend continues, we should see more and more land emerging as the blue ice melts. Even if this is true, however, it will probably be hundreds of years before the ice-river recedes from sight...and there is also the possibility that it will someday begin to advance again, as it has many times before – flowing through time.

The University of Alaska Geophysical Institute has conducted studies of the Juneau Icefield and the Mendenhall Glacier since 1997.

Geophysical Institute researchers dig a snowpit on the Juneau Icefield as part of their research.